DC LANFRANCHI

Change Management in Safety

The White Book of HSE

DC LANFRANCHI

Change Management in Safety

The White Book of HSE

Change Management in Safety

Copyright 2019

www.resiliere.com

11.436 words

First Edition

October 2019

ISBN 9781709042294

To my family and all the co-workers, seniors and organizations from whom I have learnt, and with whom I have shared my passion and my mission and to their families as well

I dedicate it as well, and especially to the experts who assist us in achieving the right evolution by means of Change Management, which in turn will lead to increasingly safer working practices.

INDEX

Prologue

This is the first time I find a short and easy reading book that summarizes so well the complex subject of Change Management applied to Safety.

I have read the 3 books previously published by Deborah Lanfranchi, and I am truly enthusiastic about her vision and mission.

She started in her first book writing about Leadership in Safety, a vital skill to produce highly efficient and effective working teams. After this, her second book taught us safety practices and their reporting areas which are three: operations, human resources and the Corporate Social Responsibility area (CSR). And in her third book she immersed us into the future of safety, showing us that it is not possible to work in the present without knowing exactly where we are ´heading for, opening the fascinating doors to the world of Singularity.

It is completely clear in what direction she is moving: towards a zero accidents working practice. Her mission is to achieve a world where everybody will work safely and will get home from work sound and safe. Who would not feel tempted to follow her and get involved in such an important mission? I feel so.

The question present in her 4 books is how we are going to incorporate Safety as a fundamental value at enterprises and organizations, and the answer is simple: making all employees aware of its importance, by means of teaching and training, until a safety gene is incorporated to our DNA with the objective of building a better and safer working environment.

Deborah told me in our long conversations that on many occasions, production managers said that doing things well took longer, and fortunately she explained it was not like this. She was also once a plant and production manager. She knows how to raise awareness and teach workers the benefits of acting safely, and that this safety saves time, guarantees better results for the business and allows working in a comfortable environment. To achieve this goal, it is clear that there is much to change in the global business culture. The dilemma we are facing now is: How do we produce changes?

In this fourth book about Change Management in Safety, she provides four clear guidelines for the design and implementation of the strategies required to achieve this; and in each book, there is an underlying key issue: Leadership, CSR, Future and Change Management. These books provide a truly clear and practical frame for these subjects, which are analyzed and dealt with in depth in her workshops.

I invite you to read this book that I have enjoyed and from which I have learnt a lot, not only from the theory included, but also through many examples. The information provided by the author is completed in her courses and training workshops, in which she includes mentoring and coaching processes, information about standards and regulations, and where she highlights the importance of globalization, good health, nutrition and attitude, among many other subjects that constitute the great value DC Lanfranchi offers to industries as regards her Safety vision and mission.

I would like to conclude this participation Deborah so warmly offered to me, wishing her personally, and this is really my intuition, that one day she is awarded the Marion Martin prize, from the National Safety Council, an extraordinary event where I met Deborah and her books.

To finish, I want to thank you Deborah for advocating for a safer and better world, with your passion and mission and conscientious work.

Safety Consultant

USA

Preface

The year 2019 is a year I will never forget. It turned out to be a vital stage and turning point in my life. I started it as safety chief in a multinational enterprise and I am finishing it, launching my consultancy and training services in safety. No doubt, this has been a year of immense learning and full of adventure. As an example of this, my four books fill me with pride and encourage me to go on working harder and harder, as they are written proof of my commitment.

To write these books, I had the enormous privilege of working with a fantastic collaborative group to whom I am grateful for their trust and commitment. I also thank the consultants and researchers, who with their vast knowledge, experience and prestige, helped me put these ideas in order.

This year, I also finished shaping Resiliere, my entrepreneurship, which was born from a question: how to offer useful safety services which really pave the way towards my Zero Accidents mission. It was then, that I finally understood that Safety has to be taught and trained as a value essential to all organizations and workers. Working safely must be transformed into a habit and a primordial need for every organization, in the same was as breathing is to our bodies.

When we refer to Change Management in Safety in the context of CSR, it is crucial to provide it with an ethical framework, emphasizing that this change will have a beneficial impact on people´s wellbeing; that is to say that after this change, people will be really better.

I strongly believe that when an organization offers to its employees a safe environment, as refers to health, wellbeing and happiness, people work better and harder. On the other hand, when people perceive this change in the organization, mainly as safety is concerned, is fake or misleading, it instantly works as a complete "stopper" of the change process proposed by the organization.

It is for this reason that I insist on the ethics framework for Safety Change Management. Without ethics, principles or truth, safety turns into a hazardous concept in itself. Because in order to work safely, we have to feel safe.

And feeling safe is enrooted in trust, without it, it cannot be achieved.

Consequently, to achieve a mission which seems to be so difficult to reach for many businessmen and leaders, we can ask ourselves the following questions: Where will I start? How will I start?

Then, I understood that in order to achieve this, I have to work in a scheme that comprises Leadership, CSR, Future and Change Management.

The central piece that puts them together is, in my opinion, Change Management, for CHANGE is the piece that bridges the gap between where we are today and where we want to get to. Now, how can I write a book on this subject that proves to be dynamic, easy, quick reading and illustrative and useful at the same time?

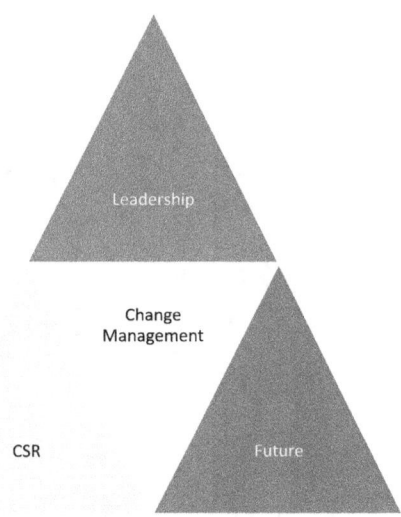

The most important idea I want to transmit in this book is that Change Management is the fundamental piece to incorporate knowledge and skills to an organization. I am truly convinced, given my corporative expertise, that Change Management must be directed by external consultants with more than 15 years experience in the subject. Why outsourced? Because seeing from the outside helps us be objective, as is required for a real process of maturing, evolving and changing.

I therefore, invite you to press the white button to start.

Introduction

The book "Change Management in Safety" is the fourth book of the Series SAFETY. I repeat what I have already expressed in the three previous books: *"my objective is to create a collection of books, which are easy to read, to remember and which can be finished quickly. I deal with different concepts in an average of 15,000/20,000 words each book, equivalent to an hour, or an hour and a half reading approximately. I intend to introduce the reader into the subject in an organized way, including the basic contents to provide managers and leaders in these areas with the essential knowledge so that they manage to carry out their duties more efficiently. In the three books, I deploy a 3 step methodology which I call ATAM, and in this order, this acronym stands for: 1. Anticipating, 2. Taking Action, 3. Measuring".*

At the exact moment I was going over this book before being published, I was reported of an incident with an injured worker who failed to do a "Lock Out Tag Out". Clearly enough, his leader was responsible for it, causing a serious problem with an OSHA inspection programmed urgently. I receive reports of this sort on a daily basis. Lately, I was also informed of an incident in Canada, where over 100 kilometer winds blew a worker off a platform. He would have fallen, hadn't he be been secured by a safety harness. Leadership in Safety is a critical factor at work.

In this book of Change Management, I start elaborating on a concept called **Distinctions**. These distinctions comprise:

- Change of perspective
- Change vs. Growth
- Resistance vs. Inspiration
- Plan vs. Program
- Information vs. Communication

ATAM in Change Management

As I have already written in my other books, **Anticipating** means to look at the future, and to anticipate, we have to be practical. For this reason, Change Management in Safety is the longest section. It includes:

- The edges of Change Management
- Why and what to change for? Co-creation of Change Vision and Value
- Building, Training and Certifying Change Management
- Appointing Change Agents

How and When will the change take place?

Co-creation of a Change Program

- Who are exposed to change? Stakeholders
- What changes? Communication
- Which is the scope of change? Organizational Impacts

In **Taking Action**, I ask "how?" questions:

- How do we get organized to adopt change? Organizational Alignment
- How do we prepare for change? Training

In **Measuring**, I elaborate on whether change should be measured or not.

- To measure or not to measure change?
- We celebrate change

I intend this book serves as a guide of the issues we should work on, a sort of check list of subjects to discuss. I did not want to go further into some of the explanations, because the details of Change Management will depend specifically on what is bound to be changed in each enterprise or organization.

It all depends on whether we are going to incorporate new personal protective elements, or if we are going to add the development of new soft skills, or if we are going to learn about a new technology.

Within new technologies, I deem it very important to point out the advantages of Virtual Training, which is further discussed in my third book, The Future of Safety.

DISTINCTIONS

Change of perspective

Before moving deeply to the universe of Change Management in Safety in particular, I would like to go over some terminology, incorporated into our language, which if observed from another angle, will offer a different perspective, very useful to achieve our goals.

Change vs. Growth

Traditionally, change has been considered a threat and it is for this, that the idea of change has always been resisted. The reason for such a resistance is likely to originate in the way in which change has always been introduced to employees (baldy introduced), and not in the concept of change itself.

Let´s consider as an example, a salary increase. I cannot remember any coworker resisting to this change. Why? Because in this case, change clearly means improvement for him.

This coworker does not ask himself: Why to me? what is going to happen to me with this change?, why does the organization want to give me a pay rise now? However, when the potential benefit a change is likely to offer is not that clear for the coworker, the answer is not positive. Questions always arise when there is a change in the organization.

Which conclusion can we draw then?   _That change must be clearly explained as contributing to personal growth, evolution and beneficial transformation for employees._

Many organizations talk about change, but only explain technological improvements (for example the incorporation of modernization) and describe it in excess. But announcing change emphasizing the particular details of new technology, does not communicate that the idea behind change is human growth.

Specifically as Safety is concerned, we must concentrate on the growth caused by safety improvements as in the case of zero accidents. This clearly represents growth and evolution from a humane perspective.

Resistance vs. Inspiration

Why both organizations and people are convinced that implementing changes generate resistance?

In the first place, due to personal past experiences. And secondly, because if you think there might be resistance ... there will certainly be. It is that simple. As we already know, our brain does not distinguish reality from mere thinking. That means that if the brain thinks so, it is exactly the same as if this was in fact happening. One popular example to demonstrate how this phenomenon operates is to concentrate on a lemon; imagine we are holding it in our hands, take a knife, cut it in halves, take one half and put it in your mouth and then squeeze its juice letting it fill your mouth and then swallow it. Do you by any chance feel your mouth producing saliva and a shivering sensation? In this experiment we clearly see the concept that the brain cannot distinguish between what is real and what is not. Then, if we think that a change will bring about resistance... it will.

. I constantly see that people program themselves and not always for good. Resistance is accepted as normal; likewise people accept that there will be accidents and injuries in the same way as they accept mathematical rules!

The first step then will be to ask ourselves:

Why do people resist change?

Here we must differentiate the various types of changes:

- Changes that due to their nature are resisted with enough ground.
- Changes that generate big and real benefits, but which are nonetheless perceived or read as a threat.

Which is the role of Change Management vis a vis each situation?

The role of Change Management will always be the never ending search to identify certain perspective that transforms change into a healthy process for the whole organization.

It is reasonable to think, for instance, that a change of the workplace location might bring about unexpected consequences for an employee, and affect the employees´ situation. The challenge will consist in creating a perspective that shows how attractive the new place will be, to compensate resistance to the change in commuting.

A change of perspective supposes to introduce change as an inspirational element, to consider it an invitation to change instead of an obligation with no apparent benefits.

Information vs. Communication

A change that is perceived as a benefit has beforehand a high percentage of success when implemented.

To understand the difference between effective communication and the mere delivery of information, we have to learn from the generation of Millennials onwards.

The generations previous to Millennials were born in an era when information was available but not wide spread; access to it was only locally. Subsequent generations were born in an era when there is Access to information everywhere, in multiple formats and languages and in big quantities. This unrestrictive access made this generations develop something called "filter", which entails developed discernment abilities, as they can separate "necessary" from "unnecessary" information.

The same principle must be applied to communications about Change Management. They must be clear, accurate, precise and clean. The essential must be withdrawn in every interpretation without distorting meaning.

Nowadays, we rely on a highly efficient resource such as the use of images, which convey messages with a high impact and make them more memorable.

All the resources must be used because audiences are varied and messages must reach every modality: visual, oral, and kinetic, among others.

Communication is a discipline, but also an art which must be crafted to suit every audience, at every particular time, with every message to be transmitted.

Today, when we think of communications, we must bear in mind that graphic design is a highly influential resource. Images accelerate the communication of messages.

The difference between the area of communications and the front of communications of a Change Program lies in the fact that the Change Program knows beforehand every landmark to be communicated, and which have to be transmitted face to face to achieve an objective.

One of the moments I remember most dearly of my time in production, was when it was decided to start working with High Performance Equipment. I was determined to announce it clearly, enthusiastically and with a vision of the future; first internally, and LATER to the Unions, who did not agree at the beginning of its implementation, because as they did not have all the information, they assumed operators might be affected negatively.

The mistake was not having talked to them beforehand. Today I can see that they should have been clearly informed from the very beginning of the changes that would be implemented, who would be affected and the benefit these changes would represent for workers. Once unions understood change scope and impact, the process went on developing and being implemented at a great speed, in a peaceful environment for all.

PART 1 ANTICIPATING

The Edges of Change Management

What is Change Management indeed?

It is anticipating, predicting, knowing beforehand where we are 'heading for, how change impacts and how to mitigate its possible negative effects. It means to prepare the scene beforehand and then work on these scenarios with cooperation to boost empathy and broaden understanding.

Change processes develop within a human context, which means that we work with a high level of complexity, with variables which are not always predictable; nevertheless, the ability to anticipate is the key to mitigate change.

We offset the high level of complexity with a systemic model that can accompany the process, and we define specific working areas, which must be programmed to mitigate Parkinson's Law (I use all the time available).

The core areas constitute the working framework for the process of change and are enhanced with Leadership and co-creation methods, Design Thinking, Journeys Design, Celebration and Rewards:

Change Management Approach

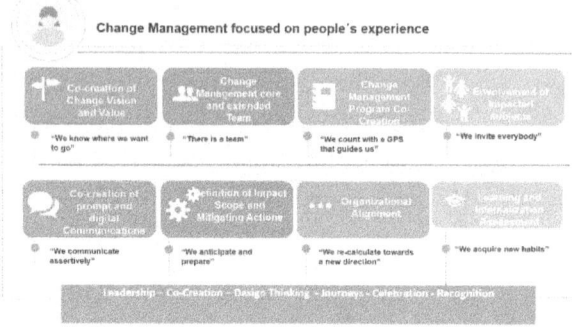

Why to change and what for? Co-creation of Change Vision and Value

Once the decision to change is made, the moment comes to go further into the reasons for such decision, so that this change is fully understood.

Helpful questions include the following:

Why to change?

Why now?

Why us?

The answers to these questions will be the axes that guide the organization towards the direction explained. At moments of hesitation or deviation, answers help us be back on the right track again; remind us of the reasons for starting the journey at the very first instance.

The result of this dimension will be to possess a clear vision of where the organization is ´heading for with the proposed change.

The search for the answers must be done with a multidisciplinary team comprising the top leaders of every area of the organization, including those not directly affected by the change.

The activity is called Co-Creation of Change Vision, and is carried out based on Design Thinking Methodology.

Is it enough to understand the reasons for changing?

It is not. It is also necessary to know the benefits of change to inspire and invite everybody to focus on results.

The identification of benefits at a high level in this instance is incorporated to Vision Co-Creation.

Is it enough to understand the reasons and benefits?

No, it is not. There is a further step, understanding impact, again at a high level, to generate empathy of the highest levels of the organization with coworkers.

Do people change?

Thinking people change may be an illusion. Thinking that a good Change Management may change people is another illusion.

Chance cannot be imposed from the outside and the unique owner of the decision to change is the person itself.

The function of Change Management is to change perception vis a vis change, and fundamentally to identify the "breaking point" people can experience to see clearly "why" to change.

What is a "breaking point"?

The breaking point is a concept used in ontological coaching, and it is the moment when something we were doing unconsciously or automatically, appears as something conscious. It implies "to become aware"; it is the moment of "aha". It is also called "Insight".

An example of a breaking point in everyday life can be the use of a light switch to turn on the light. May be one day, we touch the switch and the light does not turn on. We try again, we do not give up, our brain starts processing what is going on and then poses the question: Why is this happening?, Is the light bulb burnt?, Is there a power cut?, Is the light switch broken? We have to decide on a course of action, but which? Not any, but one which solves the problem. In a case like this, resistance to replace the light bulb or switch does not appear because we know it is the option that will solve the problem.

Similarly, Change Management deals with change needs, on many occasions defined by experts like John Kotter as the urge to change, but what really drives us to change is to recognize how beneficial this change will be.

What does "co-creating change vision" mean?

Each person determined to start a changing program may have a different understanding of the process.

Working on Co-Creation of Change Vision enables us to establish a unique understanding of the determination to change.

To do this, those who are mostly responsible for the decision to change and who represent all the affected areas of the organization are summoned and a Co-creation activity takes place.

The activity is led by a facilitator, who by means of an agenda with questions that go deeply into the motifs of the change, encourage all the participants to build a common change vision.

This exercise on agreeing is conducted with panels as shown in the figure below. Each participant can express his view to demonstrate that there will not be unique understanding of the change, and the moment will come to generate agreement and compromise.

I remember that during my job at a manufacturing plant, we decided to start working on values. I was in charge of 500 beautiful people, who I got to know individually over the 4 years I performed my duties.

To define the values of the broad team, that is to say of the 500 individuals, we conducted surveys, focus groups and dynamics so that everybody participated and were taken into account.

This experience resulted in several regent values to guide our daily actions, such as respect, team work, and solidarity, all of them essential to Change Management in matters as important and sensitive as Safety.

All the participants felt as part of this creation because they had in fact been part of them. We jointly built the framework of the values guiding our duties.

Within this scheme, one afternoon we decided to paint the bathrooms at our working sector, as they were written and deteriorated because of neglected use.

Women chose orange and men green, according to Pantone corporate colors. We painted them ourselves. We did a great job, they were beautiful. Needless to say, for long nobody graphitized or deteriorated them again, because we all felt part of this great change in our quality of life and wellbeing at the workplace.

Building, training and certifying Change Management teams

Building a Change Management team is one of the first activities to take into account. A successful change process requires an excellent Change Management team.

This team will be responsible for the all the steps relative to change management, but it will not be the only group of people involved. Its integrants will be people highly committed to the enterprise and the change proposal.

Which activities are required at this stage?

The first activity will be to gather people and afterwards provide them with training in the activities involved in a change process.

This training must be certified, so that the person in charge has written proof of its fulfillment and can repeat the change process in the future.

How many people must a Change Management Team comprise?

In alignment with the areas of Change Management, the team must have:

- A leaded expert at Change Management with facilitating skills and Agile and Design Thinking methodologies (permanent). Preferably outsourced.
- A person in charge of Communications (permanent). This person must belong to the enterprise.
- A person responsible for Training (training landmark). This person must be an expert in the matters subject of this future change.
- A person responsible for Impact and Organizational Alignment (impact analysis landmark). This person must be outsourced.
- A Graphic Designer (communications landmark). External to the organization.

Appointing Change Agents

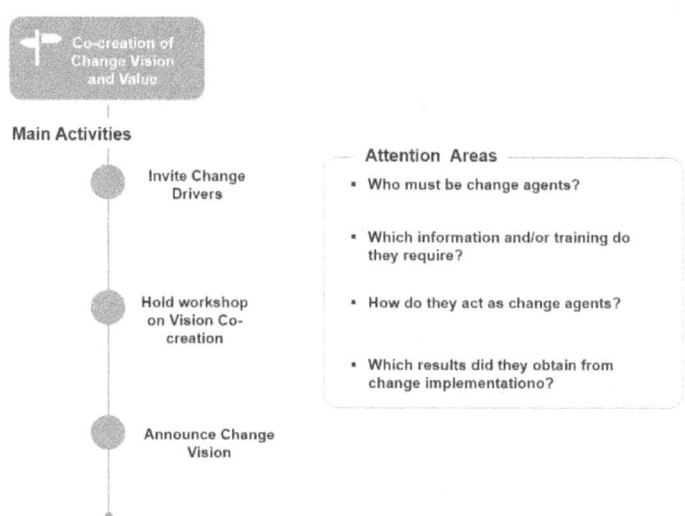

Main Activities

Invite Change Drivers

Hold workshop on Vision Co-creation

Announce Change Vision

Attention Areas

- Who must be change agents?

- Which information and/or training do they require?

- How do they act as change agents?

- Which results did they obtain from change implementatiuno?

At the onset of my professional career, I was faced with the need to transform safety culture deeply at a 3000-employee plant, but only with a HSE team of two people. It was completely impossible to reach all the employees and ensure that change would take place in a reasonable period of time and efficiently.

To carry out my duties, I relied on a team of Safety Replicators, who were nothing but Change Agents. Most of them were invited spontaneously, but always considering communications skills, empathy, interest in safety and common wellbeing and above all that the person was a real example of living safety as a value both at and out of work.

Another natural change agent for safety is the Emergency Brigade, who vocationally possesses all the characteristics mentioned before.

Is one Change Management team enough?

It is necessary, but not enough. A team focused on the Initiative of Change Management is required, because it takes the helm of the program, but to repeat the change is not the task of a 5 people team, but the joint action of multiple actors called Change Agents.

The process of change is like a journey. It starts with the dream of somebody who imagined, and perhaps some years later, transforms this dream of changing into factual implementation.

It is essential to achieve the critical mass (as in nuclear physics), that is to say the number of change agents needed to start, conduct and sustain change, in time and transcending people.

Why are they called Change Agents?

Because an Agent is an entity that renders a service. In the context of change, an agent is the one who assumes the Responsibility of communicating change. Today, people working at organizations are worried about their personal development and are trying coaching disciplines. They are trained and certified, as it represents a form a personal growth beneficial to their workplace, due to their knowledge of themselves and the empathy acquired in their training and practice.

Ideally, the agent has to act as a coach, as somebody who accompanies, who gets the best of people, who helps raise awareness, recognizes the breaking point, boosts one's morale, and who achieves all this without imposing it.

Who can be Change Agents?

Any person with a positive outlook, a mind open to new paradigms, people-oriented, recognized by his peers for the role he performs (not necessarily a leader), with the ability to communicate in a simple way, and above all, who lives safety as a value inside and outside the workplace.

Any person, who gathers these characteristics, may be a potential Change Agent for the transformation of safety culture in an organization.

According to the scope of the impact of the Change Program, the groups which require a change agent are identified considering that there must be at least one Change Agent every 30-50 people.

Leaders, due to the role they play, are naturally appointed, but they will be chosen only if they meet the characteristics mentioned before. On the contrary, in the case of a negative leader, that is to say a change Stopper, the enterprise will have to evaluate this leader's conversion.

Are we born Change Agents or do we train to be Change Agents?

Any person can be a Change Agent, but it requires training to broaden the knowledge of this potential role. To be a Change Agent, a person must know: Which are the expected results of his acts? How will they be measured?

Ideally, he should have a "journey" kit to help him in his duties, asking himself frequent questions and knowing the answers; facilitating methods and other aspects useful for the role he has to play.

To this end, he will require training to align and rectify technical vocabulary, tools, communications strategy and he will also need a Schedule of subjects and trainings to share with the rest of the staff, so that all the areas move ahead in a coordinated way.

When and where will change take place? Co-Creation of the Change Program

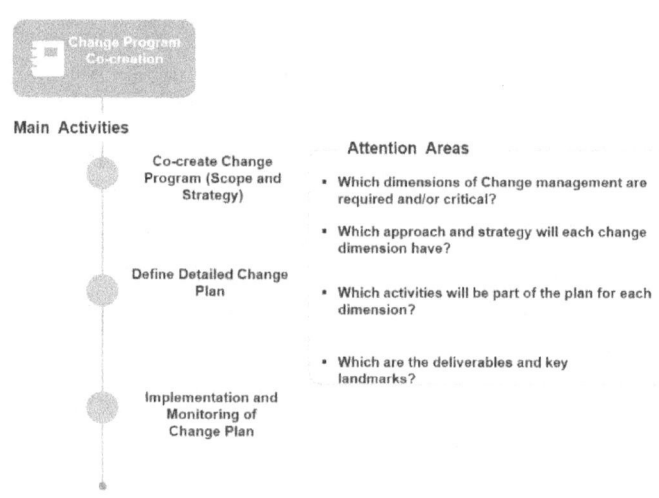

Main Activities

Attention Areas

Co-create Change Program (Scope and Strategy)

- Which dimensions of Change management are required and/or critical?

- Which approach and strategy will each change dimension have?

Define Detailed Change Plan

- Which activities will be part of the plan for each dimension?

- Which are the deliverables and key landmarks?

Implementation and Monitoring of Change Plan

Co-creation is a Change Program which makes the approach of the change plan participative, engaging multidisciplinary views and bringing it closer to those on whom change has an impact.

Who participates in the creation of the Change Program?

This invitation must be extended to a representative and at the same time to diverse groups in the organization. Diverse as regards disciplines, roles within the organization and personalities.

Disciplines: refer to the activities carried out at the organization and to the area they belong.

Role in the organization: variety in terms of leadership levels and positions held.

Personalities: variety of personalities in a working team help the activity develop softly or come to a halt. If people are mostly abstract, the idea will not be materialized. On the contrary, if there is a prevalence of skeptic realism, they will leave a blank page. A group with very similar point of views and behaviors does not generate the healthy tension required for lateral thinking.

I remember a case in which an enterprise had agreed to change the communication process for absenteeism. What had not been agreed was what each area expected from this change, so each area achieved different satisfaction levels.

Production area thought it would solve the absenteeism issue without the need to get leaders involved directly; the medical area planned to solve communications issues by means of a computing system and the area of human resources wanted to manage absenteeism from its staff administration base.

From that moment on, I really value it when <u>before defining change</u>, the organization finds a common denominator about the results expected from a change proposal, so that everybody pulls in the same direction.

How are co-creation activities carried out?

The support method used is Design Thinking, which allows articulating the cooperation of all participants, disagreeing and agreeing on ideas.

It starts with the vision of change, and they jointly define approaches for each dimension of the Change Program (stakeholders, communications, impact management, and learning) at a high level, due to the fact that new training workshops will be offered for the further study of each subject.

Which result is expected from co-creation dynamics?

The result will be to have a previously agreed approach, with general guidelines, but with a plan that contemplates real dates and committed deliverables.

The detailed plan must act as supporting material rather than as control material. If a plan is neither started nor finished, and eventually ends up being just an exhaustive control of traffic lights, the objective of the plan is completely distorted, as originally, it was conceived to indicate in which direction to move and when we have reached our destination.

Who is exposed to change? Stakeholders

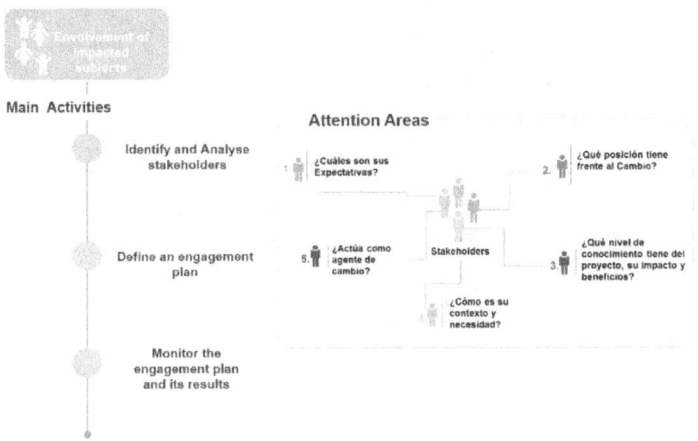

In the context of a change initiative, stakeholders are all these people or entities which will directly or indirectly be impacted by the change.

Which is the role of stakeholders?

The role assumed by stakeholders has such an influence in the change process that they can contribute to or hamper change adoption by the organization.

How do stakeholders influence the change process?

This reaction is ternary:

- They support change unconditionally.
- They resist unconditionally.
- Or they are indiferent.

It is vitally important to know stakeholders´ behavior, to be aware of the starting point and thus know if we are facing expected, resisted or neutral change.

Who are they and how are they identified?

The first great division of stakeholders we can do is between those who encourage and are accountable for change and those who are impacted (in different degrees) by it.

The fact that they encourage change does not necessarily mean that they agree, and the fact that they are impacted by it does not imply they oppose it.

A Stakeholders´ map must comprise the total number of people or entities, so we need some criteria.

We can use the hierarchical organizational structure or not, but it must include all the participants.

The internal map of the organization will have 3 groups that will work as the starting point:

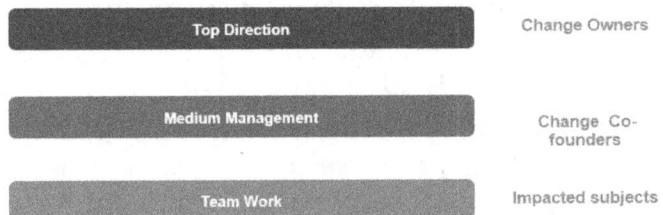

- **Change owners:** they are who make the decision to operate change, who are accountable for it and who understand in which direction to go. Ideally they assume the role of Change Agents in their working environment.
- **Change co-founders**: they are the leaders who possess the most challenging task, because on the one hand, they must understand and assimilate change as their own for their leading role, and on the other, they must accompany their teams in the transition, which automatically turns them into change agents. They constitute the inflection point because they are impacted on and they change agents simultaneously.
- **Impacted ones**: even if the whole organization is impacted, those who undergo the greatest impact are the working teams represented in co-creation instances, but do not have any direct participation in the definition of the Change Program.

As the Change Program goes further in its implementation, the different groups of stakeholders are being analyzed. It is not necessary to know the attitude of all the participants at the same time; its rhythm is set by each landmark in the Change Program.

In the first place, we need to know the situation of change owners, after this, we have to reach change co-founders and lastly we have to see to the needs of those who most require to be accompanied as they are the impacted ones.

How is stakeholders´ attitude towards change analyzed?

Whichever method is used for the interpretation of stakeholders´ perception of change, this must be face to face. Only can we have a real understanding of change perception if we can observe people thoroughly, their body language, their conversation, their thoughts and their spontaneous acts.

Every possibility of interaction with the stakeholder must have the excuse of a dialogue worthy for the person.

The meeting structure must have 3 objectives:

1- To collect information about the level of change awareness and understanding and mainly about whether its impact can be visualized (what is known is less threatening). In this instance, we must allow the person to express what this change represents both for him, personally and professionally, and for his environment.

2-to reinforce the objective of the change with clear and accurate messages, discussing the characteristics of the planned process.

3-In case of being appointed Change Agent, make use of the proper appointment instance and ensure involvement. In this way, we are not only communicating the change, but making people part of it as well.

How are stakeholders classified?

It must be noted that stakeholders´ situation must be attributed to their behavior as regards change and not to the person itself.

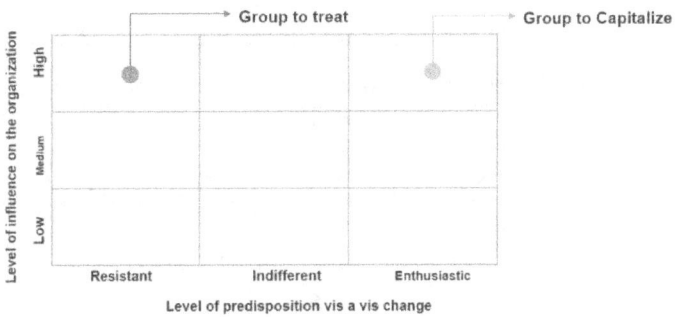

Analysis variables are 2, each including 3 states:

Variables

1) Level of influence on the organization, independently of his role (high, medium, low)
2) Level of predisposition vis a vis change (positive, neutral or negative)

Attitudes

1) **Resistant**: his behavior is hardly collaborative, resists changing, possibly due to a poor understanding and it threats perception. Actions to take: understand the cause of his anger and value his participation.

2) **Indifferent**: he ignores what is happening, neither disapproves nor participates, he is not motivated. Actions to take: understand the origin of his discouragement to act accordingly and invite him to join.

3) **Enthusiastic**: he shows enthusiasm for change; clearly understand the reasons and benefits. Actions to take: capitalize his energy to pass it on to the others.

In the steakholders´ map we will have people with various attitudes as regards the change and its influence.

We are in the presence of two main attention focuses, characterized as the positive and negative pole respectively.

- **Negative pole:** they offer high resistance and also strongly influence his co-workers.
- **Positive pole:** they show their agreement on change; they see the advantages and can inspire the others.

Let´s see an example of negative responses that may occur when we do not engage the Safety area in the initial planning of a plant location Change Management project.

This plant was moving from one location to another, and once the process was over, the management realized that it had not planned the location of the kinder area.

Bear in mind that a Kinder looks after children from 45 days to 5 years old. When the management realized that in the planning they had neither considered the location of such a particular area, nor engaged the Safety area from the beginning, they thought they had two possible options to locate it: within the car park area or in the terrace. When these two options were proposed to the Safety area, the answer was unexpected: the only alternative offered by the Safety area was not to include a Kinder area at all. There was no way, either from a legal point of view, or health or children safety point of view.

From a legal point of view, this place could not be allowed due to the building conditions. As regards health, because these location options were not healthy for a Kinder. And as regards safety, because in the event of an evacuation, it was impossible to go down the stairs various floors with 50 babies and children.

This example is the sheer evidence of how importance it is to engage the safety area from the very beginning of any change Project and/or transformation in an organization.

How can we engage both change owners and the ones suffering its impact?

The answer is empathy, understanding how they see change from their view, to contribute with a new perspective, possibly with more information. Only understanding will allow them to see beyond their limited vision of change.

What changes? Communication

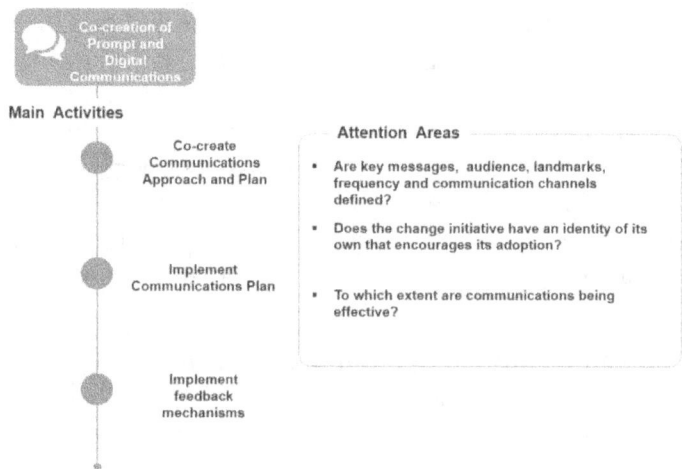

Communications offer the possibility of clarifying and understanding, of learning and of informing about the change.

Communication approaches with the most effective repercussion, use very well known formats such as communication campaigns and road shows.

- **Communication Campaigns**: they are similar to any other product or service campaign or political campaign; all the available advertising channels are used simultaneously, (billboards, web sites, signs, stickers, etc).

- **Road-shows:** they simulate a band or solo player; they have direct contact with people as they walk the organization announcing change.

Firstly, we define the approach and the communication plan and we use the co-creation modality once again to ensure that what is to be defined suits the organizational culture, or if it is disruptive, does not produce an undesired effect.

Those in charge of communications in the organization will invite people from different areas to participate in the co-creation workshop.

During the co-creation workshop on communications approach and planning, Design Thinking methodology will be used to stimulate the participants' creativity so that they can innovate in the way messages are transmitted to be reached efficiently by the different audiences.

The questions that should be made during the dynamics are:

- *Which is the best way to communicate in this culture?*
- *How can we transform culture by means of a new communication modality never used before?*
- *Which new communication support can we introduce?*
- *Which are the appropriate spaces to communicate?*
- *Who can be senders and have at the same time a high level of influence?*

The axes of these communications will be the objectives defined in the Change Vision. Nevertheless, creativity will be needed when defining how to transmit messages.

A Change Program must have a brand to be tangible and of quick reference and to be memorable. This brand will have the same design characteristics of any other product, that is to say, it will have a logo, a slogan and its own communication guidelines.

When elaborating a Communications Plan we must pose five questions:

1) What is being communicated? Message and dose (the message was defined in the vision and must be transmitted in the proper way, according to the change landmark being experienced).
2) Who is being communicated? Audience.
3) How is it being communicated? Support.
4) Who is communicating? Sender.
5) When is communication taking place? Change initiative landmarks.

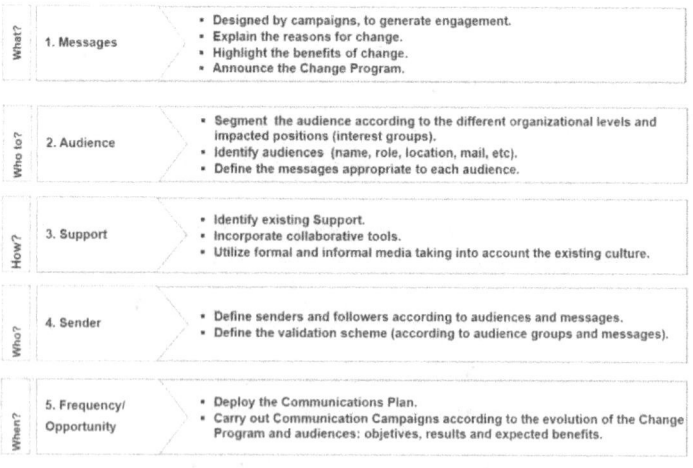

What?	1. Messages	• Designed by campaigns, to generate engagement. • Explain the reasons for change. • Highlight the benefits of change. • Announce the Change Program.
Who to?	2. Audience	• Segment the audience according to the different organizational levels and impacted positions (interest groups). • Identify audiences (name, role, location, mail, etc). • Define the messages appropriate to each audience.
How?	3. Support	• Identify existing Support. • Incorporate collaborative tools. • Utilize formal and informal media taking into account the existing culture.
Who?	4. Sender	• Define senders and followers according to audiences and messages. • Define the validation scheme (according to audience groups and messages).
When?	5. Frequency/ Opportunity	• Deploy the Communications Plan. • Carry out Communication Campaigns according to the evolution of the Change Program and audiences: objetives, results and expected benefits.

Communications Plan is the pivotal element in every communication action.

Let´s see in detail each component:

What is a campaign and how are messages developed?

The campaign anticipates every key landmark of the Change Program and must be present before, during and after the landmark.

The following example will show the due care we have to take of communications messages, audiences, moments and support in a determined landmark.

The campaign instances and characteristics are detailed below:

Before Training

1. Communicating the Training Plan to the organization executives.

Objective: to inform the impact training will have on the working day of coworkers to the highest levels of the organization, who must be informed to provide support and at the same time encourage training.

Audience: executives and operation chiefs.

Support: communications will be delivered in person to enable the exchange of ideas with those who have the vision of all the operation, and if necessary, adjustments will be made to the training program.

---

2. **Communicating the leaders that the people in their area will be invited to train, and for this reason they will have to define a replacement or contingency scheme to make up for their absence.**

Objective: **to** achieve a high level of engagement on the part of those responsible for the Training Program areas, attending to their needs and letting them carry out the proper organization of their people.

Audience: area leaders.

Support: communication will be face to face to allow fluid interaction.

3. **Communicating coworkers that they will be invited to training workshops.**

Objective: to tell future trainees the impact the training program will have on their working day, while stressing the importance of their attendance.

Audience: coworkers who should receive training.

Support: communication will be carried out via:

a. an individual mail with high visual contents, which invites them to receive training and motivates attendance, and including all the data required to participate: place, date, subject matter and duration.

b. Physical boards and digital boards, if available, and the enterprise internal page.

c. Flyers distributed by areas with the general training plan.

While Training

4. **Communicating in detail every interested group, when they will be trained, the syllabus, location and trainer.**

Objective: reminder of attendance to the next stage of the training.
Audience: collaborators who were appointed to be trained.
Support: personalized mail will be used.

5. **Communicating deviations or agenda changes.**

Objective: inform any change in the training agenda.
Audience: collaborators appointed to be trained.
Support: personalized mail will be used.

After training

6. **Communicating the results of the communication: number of people present, people absent, training hours, and results of certifications.**

Objective: inform the results obtained in the training and give formal recognition to trainees and trainers.
Audience: the whole organization.

Support:
a. an individual mail.
b. Physical boards and digital boards, if available, the enterprise internal web page.

How are audiences managed?

The segmentation of communication audiences is very similar to the one used by stakeholders: they are categorized as owners and change drivers and audiences impacted by change.

Messages are well differentiated because each audience requires different information and must be elaborated to suit each group.

- **Audience owner and driver of change:** it receives communication referring to the advances and results of the Change program as well as, information as regards difficulties and risks. On some occasions, it can be responsible for sending the messages.

- **Audience impacted by change:** it receives communications referred to change benefits and relevant landmarks.

Communications management has other adjusting variables that include: follow-up, validation, approval, pieces design and choice of appropriate support.

What kind of support can be used in communications management?

Every support available in the organization or defined by the Change program, which is chosen according to the messages. The great number of options makes it possible to count with more interesting communication, being graphic and visual communication the ones that generate greater impact.

The support existing at the organization is used, but at the same time, change initiatives are an excellent excuse to incorporate new kinds of support.

This variety ranges from mail, digital board, physical board, flyer, screensaver, stickers, diptychs, triptychs, info graphs, giant graphs and banners to handwritten personalized messages.

Who are the senders?

The definition of senders is agreed on when defining the communications plan. Each message will determine who is accountable for its implementation.

- **Top directives:** responsible for communications launching the Change program, invitation and engagement, recognition and celebration.
- **Area leaders:** responsible for communications related to the instances of the Change Program that have an impact on their working team.

A prompt validation process must be defined. The definition of senders is closely related to the validation program: the more hierarchical the level, the more complicated the validation level is, and greater is the risk of not being able to send communications within the required terms. People in charge of validating contents must undertake to be available and to hand in the contents reviewed and approved within the agreed terms.

How often are communications released?

Communication moments are established by the Change program. Nonetheless, its intensity must be defined according to the change initiative and the organization culture.

To avoid over communicating, the intensity of communications will be increased as the Change Program moves on.

Communicating campaigns are necessary, but walking the organization and announcing change via communication actions on the part of recognized leaders is fundamental.

How is a two- way communication channel generated?

So far communications management was introduced with a single direction, only as one way channel. Creating communication double direction channels is directly proportional to the level of involvement to be achieved. Today, technology offers many alternatives: chat, collaborative platforms, applications, etc.. With an organic view, we can make use of boards and physical or virtual spaces, where collaborators can state their ideas, proposals, opinions, offer feedback, etc. In short, where they can express themselves.

Which is the scope of change?
Organizational impact

Impact is the unbalancing impact of change which moves the organization out of it axe.

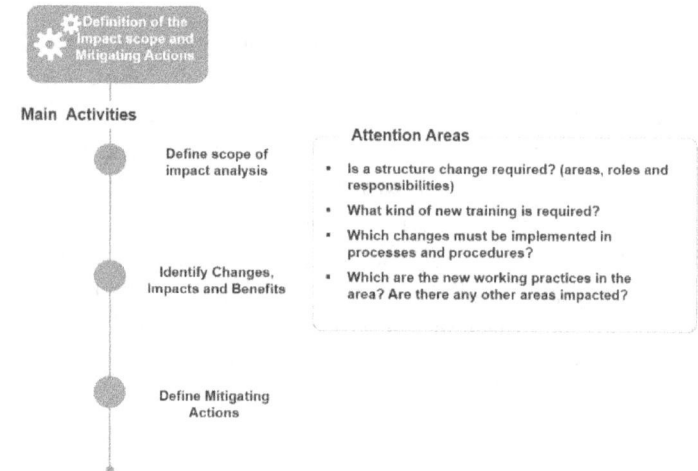

By means of the impact analysis we manage to mitigate the implications of change, and it represents one of the most powerful tools available to anticipate the negative effects of change.

What is the difference between change and impact?

HUGE, because change is what we introduced as new, as modified and impact is the consequence of this change.

It is a very common mistake to describe impact as if it were the change itself.

Change initiatives go through the organization as diminutive changes which have to be analyzed to identify what kind of impact they will produce.

For example, the change of a technological tool affects the way the organization works, we know this and we understand it. What we do not know is the way in which it will impact on each area and individual in his daily activities.

In this example we ask ourselves:

- **Which is the change?** The most evident change is the use of the new screen of the system to be implemented.
- **Which is the impact?** It is the time it will take the collaborator to use the new screen for the first time, due to the adjustment period he will have to overcome.

How do we mitigate the consequences of impact?

We do it by means of the simulation of scenarios that will be generated by impact, and we define mitigating actions to compensate them.

Due to the fact that impact is perceived as something negative, it is highly advisable to analyze and identify in parallel the benefits generated by the change initiative.

How do we analyze change impact and define mitigating actions?

In the first place, we have to limit the area of analysis to be able to see it all, and then we carry it out from four key dimensions which ensure us that we have made an integral evaluation of the organization.

The dimensions are: processes, people, organizational structure and technology.

Processes	People	Organizational Structure	Technology
Activities Change:	Role and Responsibilities Change:	Areas Change:	Access changes:
• Elimination/ incorporation of activities	• New Skill	• New	• New permits
• Duration	• New Incorporations	• Different	
• Automation		• Obsolete	

It is interesting to observe how the domino effect is produced among the different dimensions when a change is introduced. Let´s assume for example, that the change affects processes, roles and responsibilities immediately change; every area undergoes modifications and technology will have to be adjusted to the new definition of organizational structure (example: approval schemes, etc.)

On one occasion, an enterprise undergoing Change Management was not aware of the need to evaluate the consequences of a specific change with a 360 degree view of the organization, nor did it have an analysis of the whole ecosystem with which it interacted.

What happened specifically was that after a row of accidents originated by the wrong use of cutters or the use of inappropriate ones, the organization decided to change them. To this end, the safety team together with a supervisor chose new cutters safer for the tasks carried out at the different sectors of the plant.

Shortly afterwards, what happened was that for another area of the manufacturing process, where the cutter was hardly ever used, the new cutter was not useful because it could not open a rigid band. Workers complained and it was only then that they understood the mistake they had made in the analysis.

This problem had not been anticipated as this operation was not included in the risk matrix, due to the sporadic use of this element and because these workers had never been consulted about the use of the cutter.

From then on, it was clearly demonstrated how fundamental it was to make all the relevant questions to the workers, who are the ones who really know what they need to perform their tasks more efficiently and more safely. This is the 360 degrees assessment. As this 350 degree assessment had been overlooked, workers were never consulted about the use they did of this element. Today, it is impossible to think about implementing a change without the participation of the main actors, who are the ones that really know about daily operations.

If we can promptly identify what changes only, we will be able to speed up the process of identifying benefits. Each organization will be able to recognize the incorporation of other analyses for a better implementation of analysis methods such as potential risks and change implementation barriers.

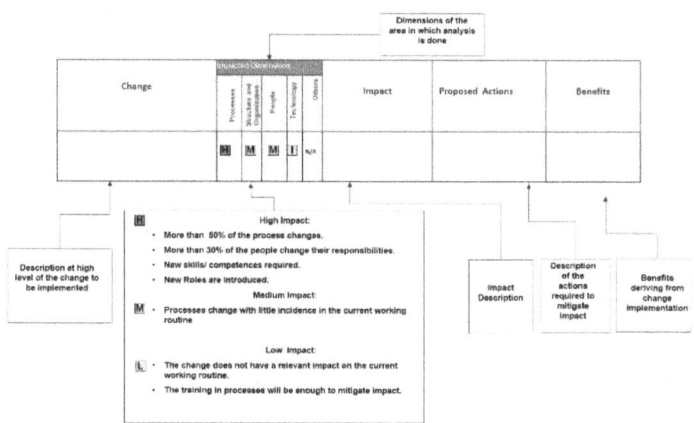

Dimensions of the Impact Analysis

Dimension	Change	Impact	Benefit
Processes	Activities that change or modify or eliminate .	• Need to train new skills • Incorporation of temporary or permanent staff • Relocation and support of people whose tasks will be eliminated	Process improvement
People	New Roles or responsibiti es	• Need to train new skills Incorporate temporary or permanent staff	Development of people
Organization al structure	• New areas • Change of areas	• Need of a new physical place or setting	Optimization
Technology	• New technolog y	• Integration to current systems	Updating

The analysis performance must be carried out in a co-creative way, inviting actors who are key representatives of the affected areas. Journeys will be used as analysis bases. "Journey" is the name given to the route followed by the activities of the process, but including the emotions of workers impacted on, who are a fundamental piece of the analysis.

It is articulated by means of workshops based on journeys construction methods. Each journey comprises a document with its corresponding benefit and the definition of a mitigation action for the identified impact.

Change can or must start in Safety?

In a multinational company where I used to work, a profound global change was being carried out. The plan stated that firstly, the change would be implemented at the production level and then it would be extended to the rest of the business, but it was never mentioned that the objective of the change was to improve safety practices.

When the first changes started to happen, the company was faced with resistance on the part of workers, who demonstrated a clear ignorance of the facts which motivated change and of their new roles arising from this change. This situation was due to the lack of a clear communication program about the change.

Analyzing the situation, leaders and consultants finally understood that change would have been better accepted, understood and accompanied, had it started communicating that change was due to the intention of the company of improving safety features. The reason was clear: had it been clearly communicated, nobody would have opposed such an amazing transformation which benefited health and safety at work and eventually people´s wellbeing. From this example we can conclude that, as far as Change Management is concerned, safety is a catalyze or "enabler" for a successful process of change and transformation.

PART 2 TAKING ACTION

How do we get organized to adopt change? Organizational Alignment

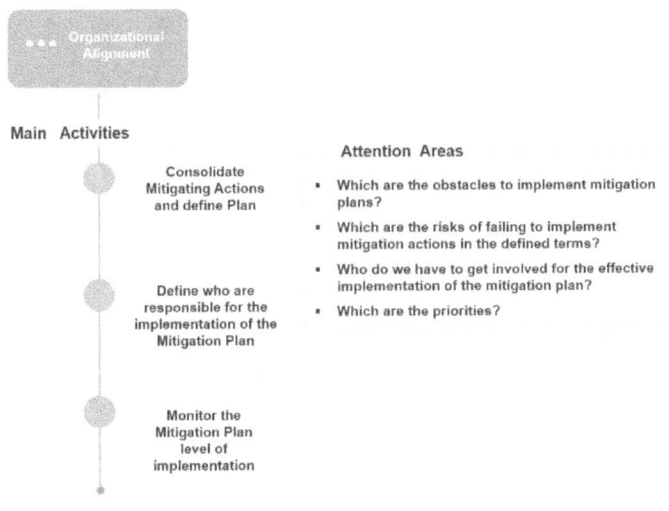

After carrying out the impact analysis, and mitigating actions, we consolidate all the plan actions to be implemented.

Actions must be prioritized to define the relevance level the Change Program has, as regards their implementation. Due to the urgency of the implementation of mitigating actions, a committee including the change owner is formed to appoint responsible champions of the Mitigation Plan. Alignment allows everything to go back to its original place;

processes, roles, structure and knowledge finally are where they should be.

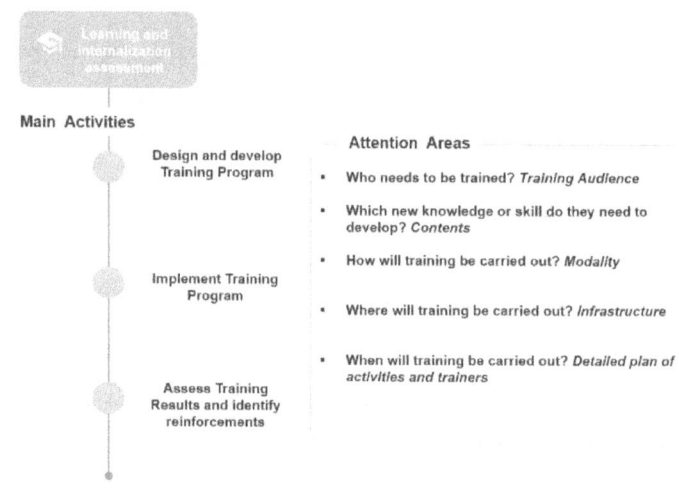

Learning and internalization assessment

Main Activities

Design and develop Training Program

Implement Training Program

Assess Training Results and identify reinforcements

Attention Areas

- Who needs to be trained? *Training Audience*
- Which new knowledge or skill do they need to develop? *Contents*
- How will training be carried out? *Modality*
- Where will training be carried out? *Infrastructure*
- When will training be carried out? *Detailed plan of activities and trainers*

How do we prepare for change?

Teaching and Training

Training is the mitigating action by excellence for most impact deriving from change.

The first step will be to invite the training experts and coworkers of the organization who represent the different areas to attend the workshop of co-creation of the Training Program.

In this workshop, the best training options will be created, with a critical view of the current training schemes in order to propose new effective modalities, updated to suit the new learning reality.

The difference between Learning and Training should be noted. Training **supposes** the short term and is generally oriented to help people develop skills and abilities for a specific job.   While learning is focused on providing knowledge, training focuses on developing skills.

Coworkers´ preparation is essential, and must be considered in the following aspects:

The person

From the point of view of safety, we consider that a person to carry out his task efficiently and safely must:

1) KNOW how to do it based on his knowledge. Due to these, training courses and specializations must be encouraged.
2) BE ABLE TO do it as regards his psychophysical aptitudes, for which he must receive training for example on how to improve his fine motor abilities, decision making and to enhance his risk perception, to name just a few.
3) AND WANT TO do it, as regards his behavioral aptitudes, which are more easily modified for they are shaped by organizational culture and the cohabitation code that the institution builds by means of rules, procedures, penalties, recognitions and rewards.   Definitely, what Dupont company calls "Operative Discipline".

Due to this, the questions we must answer about people to start a process of Change Management are:

4) Do they know what they have to do? Knowledge gaps they have to overcome with training.

5) Can they do what they are asked to do? Capability gaps they can overcome with knowledge and motivation.

6) Do they want to do what they are asked to do? Motivation gap they can overcome noticing the positive aspects of their job.

I remember an occasion on which a manager was very concerned and asked me about an area chief who he wanted to relocate, because his performance in safety terms was not acceptable.

In order to be of assistance, we went through the 3 questions mentioned before, in the following way:

- **Didn´t this person have the knowledge? It was necessary to know if he had received the education and training required.**
- **This person had never done this job before but could he do it?** This meant that he had the knowledge, but needed support and training in the skills required to perform his job.
- **This person used to do his job well and at that time he didn´t?** If that was the circumstance, it was a motivation problem and it had to be dealt with further.

Developing motivation is not the same as transmitting knowledge and is also different from training skills. To develop motivation, it is necessary to understand what the other knows, what the other feels and what the other needs. Developing motivation in safety means to connect with feelings, with the family, with the wish of a better life; in fact, it means to connect with life.

Training does not start and finish with the transference of technical knowledge.

Training does not start and finish with the transference of technical knowledge only. For an effective and efficient training, an integral preparation is required, including motivation and the person´s growing potential.

There are key elements of Training that need to be taken into account to define the Program.

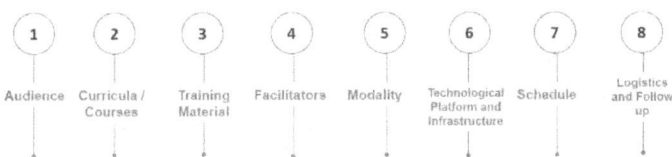

Let´s read each of them:

- **Audience**: comprises all workers who require learning. They are classified according to the use they make of the acquired knowledge. The greater the impact of change, the bigger the intensity of training. There are workers whose tasks are hardly impacted, and thus do not require practice.
- In cases in which change transforms their tasks, it is highly advisable to offer practical training through simulation. Even if the classroom resource is always available, these cases require practical knowledge not only theoretical.
- **Curricula/courses:** refer to contents structure, subjects and topics to be developed.

- **Training Material:** does not only include the manual, but also any other kind of support which contributes to understanding, such as flyers, triptychs, diptychs, digital material and virtual or augmented reality.
- **Facilitators**: will be responsible for transmitting contents, it is highly advisable that they are part of the organization, that they are trained as instructors and that they receive coaching from an expert consultant to become facilitators.
- **Modality:** refers to the options by means of which contents will be presented. Its definition will be part of the co-creation activities in order to use the modality that best suits the required contents: attendance to classrooms, virtual classrooms, and remote classrooms.
- **Technological Platform:** it will be necessary to count with a platform independent from the one used for the operation to avoid data loss.
- **Schedule:** workers´ preparation must have a time window of 2 weeks previous to the date set for the completion of the Change Program implementation.
  Logistics and Follow-up: include all the supporting activities which accompany training. I will list some of them:
  - Training rooms.
  - Attendance record.
  - Logistics and meals provision.
  - Contingency plans.
  - If computers are used, computers rooms, desks, networks and mainly, working sessions.
  - In the case of training in personal protective elements, their availability during training.
  - Plan of working sites.

o Timetable characteristics, let´s bear in mind that logistics is 24x7.

Training must include an evaluation of the level of understanding and learning, as well as of the contents and trainers, in order to introduce improvements in future trainings.

The importance of new technologies

Virtual reality, especially when combined with artificial intelligence enables us to potentialize traditional education.

I included here a photograph that I published in my book about the future. Here I want to show that by wearing a virtual reality helmet, the experiences of electric risk was stimulated while I was in front of an electric board.

When artificial intelligence and virtual reality converge, remote education becomes easier because we can find teachers and finance specific stimulation programs for our plants, considering their plans. It is about quality and quantity on demand. We can train at any time, at any place.

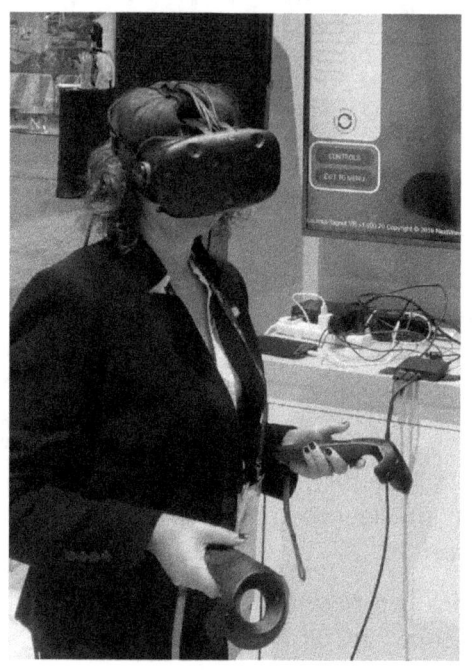

I read on the internet the following: *"10,000 out of the 1.2 million employees at Wal-Mart have undergone skills tests based on virtual reality. Learning modules that once took from 35 to 45 minutes now take from 3 to 5. The company intends to train 1 million employees wearing headsets Oculus VR by the end of this year. The initial cost of the VR headsets will eventually be recovered in the form of labor efficiency".*

PART 3 MEASURING

To measure or not to measure change?

Yes, we have to measure, measuring enables us to understand the adoption level the organization had as regards the changes proposed for implementation.

The way to make it tangible is through the definition of quantitative variables, such as trained coworkers, feedback levels of received communications and qualitative opinions of those who acted as Change Agents.

Multi-location Measurements

We no longer work at a single place; instead we do it interacting in multiple spaces and also in multiple realities, as we cohabit with real change agents and agents on line. This coexistence can be enhanced with augmented reality, virtual reality and artificial intelligence.

The advantage of these technologies is that they allow us to establish thousands of measurements and metrics combined in real time. If everything takes place in a programmable environment, then statistics of use can be exported to a spreadsheet and all sorts of risk assessments can be elaborated there.

We will be able to design clothing and personal protective elements in a virtual scenario before wearing them in the real world. The concept of multi-location now comprises virtual location.

Each time we add digital components, we are also adding the possibility of measurements immensely valuable, which can be combined in risk analyses, and in real time!

We celebrate change

Eventually, after overcoming many setbacks, the moment came to celebrate, recognize and be grateful.

The road to change is not easy for anybody, but is more challenging for those who are at their workplace every day. They are the ones who carry out tasks, leave behind old practices and incorporate new ones without neglecting their duties.

Celebrations are not only a reception on a launching day, on the contrary they must be a continuous, evolving and constant habit. I urge you to implement this practice which motivates while empowering workers.

Change is really worthy if it contributes to make every day in the life of workers more entertaining, happy and valuable. When this takes place, change is unavoidable and those who will more enthusiastically advocate for it, will be the ones on whom it impacts directly and positively.

In the same way as we emphasize the importance of creating safe and healthy habits among our employees, it is also vital that we generate celebration habits for small and big achievements within organizational leadership.

The theory of safety based on behavior states that the recognition for safe behavior is five times more powerful than the effect of disciplinary measures for unsafe behavior. This is an indisputable conclusion about the effect recognition has in the brain in particular and in human emotions in general. If we count with this information, why not being also change agents to promote actions that generate enhanced health, happiness and wellbeing in our organizations? Let´s celebrate then!

PART 4 AFTERTHOUGHTS & REFLEXIONS

A crisis occurs when the old does not finish dying and when the new is not yet born

Berlot Brecht

Why do crises happen and the end? April 2020 Reflections

All crises happen and then they go away because everything does, and all that remains is what we were able to build in both times of peace and in times **of turbulence..**

When we talk about issues of Safety it does not escape that law. Today more than ever, as leaders, we must be alert and vigilant about the health and safety of our people. And when I say our people, I mean our family, friends, work teams, neighbors, collaborators and ourselves.

Self-leadership in terms of safety has become, in this time of pandemic, an even more fundamental issue than it already was before. To not perceive the risk that   surrounds us today without the seriousness and social consciousness necessary is a trap in which we can fall from considering the enemy an "invisible" one, as are many of the risks that surround us on our daily life. And that's why we let our guard down, because we don't see it or we get used to living with it, or we create to ourselves a false sense of safety where we make ourselves believe that things are ok, things are going well or things are getting better on their own.

And so, we allow ourselves to loosen up our defenses and common sense in the way of building the awareness of safety and health. That´s a journey that does not have a happy ending.

To achieve some sort of happy ending in post of the caring of the most precious asset that we have, which is our life, we must work permanently in the search for continuous improvement.

Improving our organizational processes, improving our safety management processes, improving the safety culture and above all things improving our own skills as leaders and even more so as people, it is critical to be able to face the challenges of this new world that is brewing. It is true that every crisis brings new opportunities in return, but in order to take them you must be prepared.

We have in our hands the opportunity and responsibility to exercise our **visible leadership** in front of our people. We have the non-delegable responsibility for the caring of others that we implicitly accept when we claim to be their leader, their guide. The commitment then is to get into action. Let's do it!

Today more than ever our people need committed leaders, with courage and humanitarian skills and heart, ensuring the safety and health of their teams. Let's seize this moment. What we are living today is unique and historic. Let us work up to the circumstances and the needs of our people.

Many companies have the BCP -Business Continuity Plan- with crisis committees and emergency preparations. Such committees formulate simulations and tests of abnormal situations but under normal conditions. However, for the times that we live nowadays no one was prepared, and it is even more complex, since most people had not even thought about it.

Well, let me rephrase that last bit. The truth is that yes, many people in universities and government agencies thought about a scenario as the one we are living today. It was also thought by many creators of science fiction movies, but as something that can only happen in our imagination. Some lateral thinkers like Bill Gates talked about it in 2015, but he was taken as an exaggerated idea or very advanced in time.

And yet here we are, starting the year 2020 in which the worst pandemic in history has been declared. A microscopic entity, a virus, a Coronavirus, that scientists are still discussing whether a virus is a living being or not. And in the midst of all that discussion the virus is killing us and is killing the world, literally and not so literally speaking.

The worst deaths are the real ones, the deaths of thousands of real people, no doubt about it. However, before the situation normalizes, this global crisis will shake us with its aftershocks, as all great earthquakes do, that will be many and there will be plenty of other types of deaths.

Many things and dynamics will definitely change. Money and the relationship with money will change. Perhaps cryptocurrencies and virtual money will be kicked out of the investment portfolios of a few enlightened ones and get pushed into everyone's daily life in the rise of the world of the new order. That would prevent, for example, queues at banks, pharmacies and supermarkets.

Major and important changes in cleaning and hygiene rituals, both in businesses and homes, are changing and that change has come to stay.

And without a doubt what has changed forever is work. Everyone's work. Many of us resisted the virtual world, because the one-on-one contact was part of what we wanted to feel, we wanted to feel close to other people. Work will forever change as will many other experiences that we believed it could only be done by being physically present. Today´s reality confronts us with many of our limiting beliefs, many of which we live and experience in so many areas of our lives, and that are also to be found the area of safety.

The year 2020 came and from one day to the next, suddenly and in a blink of an eye, without asking permission or giving us time to organize anything, our home became our office. But not only our house was transformed into the office of all those who work in each home, but it was also transformed into a school, a kindergarten, a gym, a restaurant, a hair salon, everything!!! Our homes have been transformed into everything 7 days a week 24 hours a day. Many of us were brought a computer to our house the very next day and our children were no longer welcomed at their schools. At that very moment the real change began. Because in addition it was no longer allowed, nor was it safe, to leave our houses.

To work well we must have a space that accommodates all our needs and requirements of all kinds. That is why it is extremely important to have a safe, healthy and happy design and equipment for the home office workspace.

Let's look at several important points that we can contribute from the Safety area for this new stage of a home office:

Ergonomics

Ergonomics is the science that adapts the job to the activity and the person who develops it and not the other way around. Since the industrial revolution was always prioritized or played with the incredible adaptability that human beings have to everything..

Even viruses!! And then tasks arose at very high temperatures, with heavy materials, with forced positions based on that imposed carrot, sometimes wonderful, that drives us to believe that human beings can do everything.

However, revolutionary thinkers emerged and with them the concept of Corporate Social Responsibility (CSR), which made us see and understand that not everything should be done, even if possible.

Although the term was first coined in 1857, in Poland, it was in the 1970s, well into the 20th century, that ergonomics took its relevance as we know it today.

The objectives of ergonomics are:

- To reduce or eliminate occupational risks, labor accidents and diseases.
- To decrease physical, psychophysical and mental fatigue
- To increasing the efficiency of productive activities

The bases and laws of ergonomy are already established and tested, and by them it is understood that for the safe, healthy and happy design of the workplace, the following four principles must be taken into account:

1. **The locative risks**: i.e. the risks inherent in the workplace itself.
   o   Detect loose objects
   o   Don't stan on tables or chairs to reach objects
   o   Close drawers to avoid bumps and stumbles, among others possible accidents.

2. **Ergonomic risks:** are those that arise when we neglect our bodies position during the execution of our work.

   o   Take into consideration the importance of lighting, the more natural and direct, the better. Avoid irritating and uncomfortable reflections on work surfaces.

o Arrange the more frequently used objects nearby and at your fingertips, and the least frequently used objects in a wider radius distance.

o Keep your elbows and knees always at 90 degrees in relationship to the body axis.

o Keep your feet firmly resting on the ground. To accomplish this, you can use shoe boxes, wooden drawers or cushions.

o Keep the dorsal spine comfortably supported by the back of the chair (yes, you have to choose a chair with backrest), if you can't get your back to rest comfortably on the backrest you should try using some cushions.

3. **Electrical risks:** these are more common than we think and perhaps the least measured.

o To avoid them keep the cables neat and out of the way to avoid tripping and falling over

o Don't place glasses with water or infusions in places where you might inadvertently dump them on your keyboard or computer

- o   Don´t overload plugs
- o   Don´t use extenders that are not approved by a licensed electrician
- o   Electrical installation, both at work, and at home must have differential protection and circuit breakers

4. **Psychosocial risks:** These have become more relevant today than ever before due to periods of obligatory social isolation, but it goes without saying that home office work has always been and always will be a challenge, especially regarding the maintenance of fluid and healthy interpersonal relationships, even more so during these special times.

Building happiness

In addition to everything mentioned above, it is important to have a work environment that makes us happy, that provides us with well-being, comfort and joy. And so, we must build that environment. Yes, build it. Because the environment that leads to happiness is built in all areas of life.

To build happiness we will use the science of positive psychology. This science ensures that 50% of the level of our happiness is due to our hereditary base, 10% is due to the circumstances and the remaining 40% is due to the intentional activities we do in our lives. That is why we can affirm that happiness is built, it is a choice that we can make. Science shows that we have a 40% of "space" to shape our lives with nice thoughts and feelings that come from doing things we like and makes us feel good.

Healthy Routines

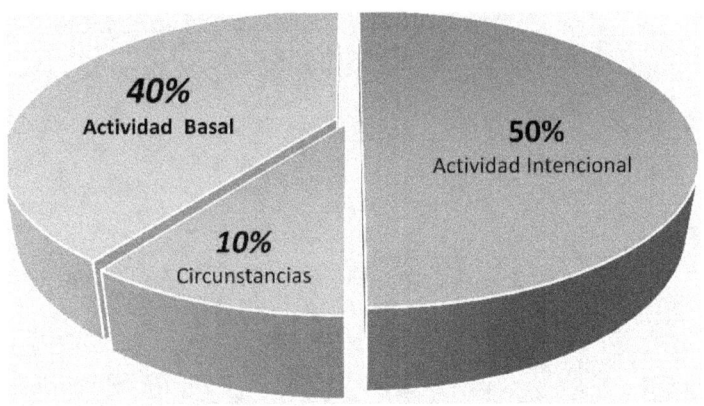

Another important part of our home office routine is being able to make **active micro breaks.**

Active micro active breaks or micro active pauses serve us to make some muscle stretching exercises. It is necessary to take them every certain interval of time to help prevent diseases typical of those who spend many hours in little or no movement and can expose us to two specific dangerous risks: sedentary lifestyle and non-ergonomic body posture.

Unlike an accident, which is a sudden and unexpected event, occupational diseases manifest over time, as they are the consequence of cumulative and gradual exposure to poor work habits.

Obviously, for an occupational disease to develop, an individual must also have a basal load of predisposition. There are people who, at the same stimulus, do not develop a disease that others do, or develop it later, or it manifests itself with less severity.

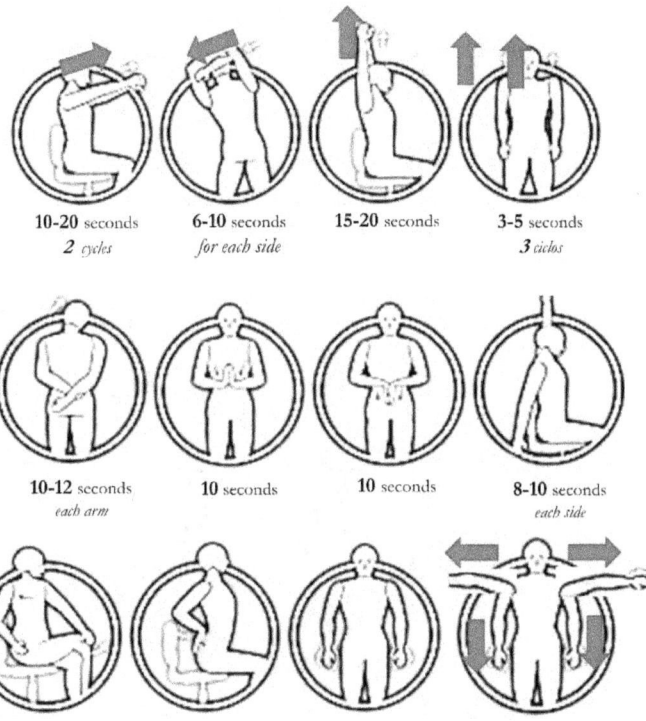

10-20 seconds
2 cycles

6-10 seconds
for each side

15-20 seconds

3-5 seconds
3 ciclos

10-12 seconds
each arm

10 seconds

10 seconds

8-10 seconds
each side

8-10 seconds
each side

10-15 seconds
2 cycles

6-20 seconds
shaking hands

10-20 seconds
strech the arms

Performing 3 to 5 minutes of micro-breaks every three to four hours of work will

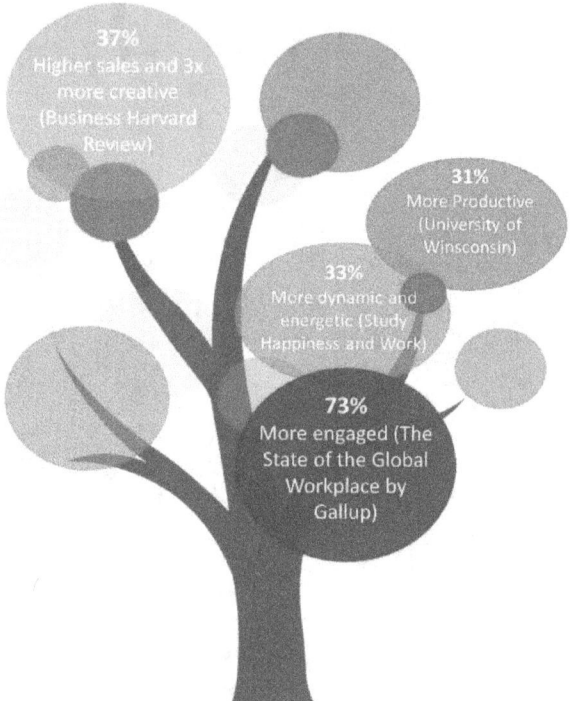

improve not only our body, but also our attention, our mood and our mood. Staying on the move is extremely important to our psychophysical balance.

Happy companies are those that make sure their employees are happy, because organizations are the sum of all their employees.

Human beings are a unique and indivisible entity, and from ontological coherence we are what we say, and that transforms us into what we do. To be happy we must think, feel, say and do in a coherent way, that is, harmoniously.

It´s a good thing that many years ago we got rid of a very limiting belief that said that personal problems could be left at the entrance door of the office, or at the entrance gate of the factory, or in the entrance hall of the house next to the umbrella and the coat. Today we know that this is not the case anymore because it´s neither possible nor healthy.

My reflection in this annex is purely to invite you to walk this path of coherence and allow emotions to emerge in the area that needs to emerge, thus avoiding much stress.

Stress

The workload that the employee believes they cannot handle is called job stress, and nowadays it is the second most important and frequent cause of work sick leave followed by musculoskeletal causes.

What is striking is that both causes, musculoskeletal and emotional, are affecting all industries and organizations without distinction of which category they belong to.

Today, in the midst of the Coronavirus pandemic spreading through our streets, the two most prevalent causes of sick leave are in full bloom.

We have the feeling that our bodies and minds have been imprisoned by an invisible enemy against which we cannot fight. But it's not like that. We can fight against it from what we know, which is to take care of our body with the tools that give us safety, hygiene, ergonomics and above all our minds, with the tools of positive psychology, creativity and mental flexibility.

.

Happy and healthy employees build happy and healthy organizations. Healthy organizations have been proven to be more efficient and profitable business. Therefore, it will be the organizations that understand this important concept and invest on it all their resources (human, capital, know-how, lateral thinking, etc.) the ones that will survive to these or any other time of crisis.

Epilogue

We have reached the end of my fourth book in this first year.

On many occasions I was tempted to write longer pages, but I know that lengthy books are eventually abandoned. In my experience, literature about specific matters should be understandable, brief, practical and quick paced. I also consider that it should be available in digital and printed formats; it should be in my mother tongue (Spanish) and at least in English because it is the language of business all over the world.

I wanted this book on Change Management in Safety to comprise a list of topics to guide, as a sort of check list, readers in the mental and practical journey of Management.

At my workshops, I encourage you to develop Excel spreadsheets containing the items discussed in this fourth book, as well as those in the previous books and to elaborate really profound information metrics. Each organization will choose to emphasize there, its particular matters. Change Management implies to understand that standard spreadsheets are useful to plan clothing sizes, but to manage change we have to explain in depth, for example, how these sizes are constructed.

As always, I hope I have been helpful with my contribution of real and practical knowledge and above all to inspire you to go on investigating the subject. I invite you to write to me at debbie@resiliere.com and send me comments, proposals, ideas and questions and to visit my site www.resiliere.comParkinson to keep informed about my workshops and educational programs.

I hope to find you there, on the other side, in my next books. I am very grateful for your support.

Deborah C. Lanfranchi

Experienced leader in HSE and Manufacturing with vast working background in multinational companies. Expert in Corporate Management Systems, Risk Management and ISO Standards. Certified Professional Coaching, Life Coaching. Chemical Engineer. UTN/UBA. Analyst of Safety Behavior Metrics. Associated to CPIQ.

Education

Chemical Engineering. Universidad Tecnológica Nacional – (Public University of Technology), Argentina.

Professional Ontolological Coaching.

Specialized in Industrial Safety and Hygiene – Universidad de Buenos Aires (University of Buenos Aires) – Argentina.

Former teacher at the Post-graduate course in Safety and Hygiene Specialization, School of Exact and Natural Sciences - Universidad de Buenos Aires (University of Buenos Aires) - Argentina.

NOTES

_____

_____

_____

_____

_____

_____

_____

_____

_____

_____

_____

_____

NOTES

NOTES

_____

_____

_____

_____

_____

_____

_____

_____

_____

_____

_____

_____

NOTES

NOTES

NOTES